THE WOODS

written and photographed
by
Mia Coulton

Here is a rabbit.

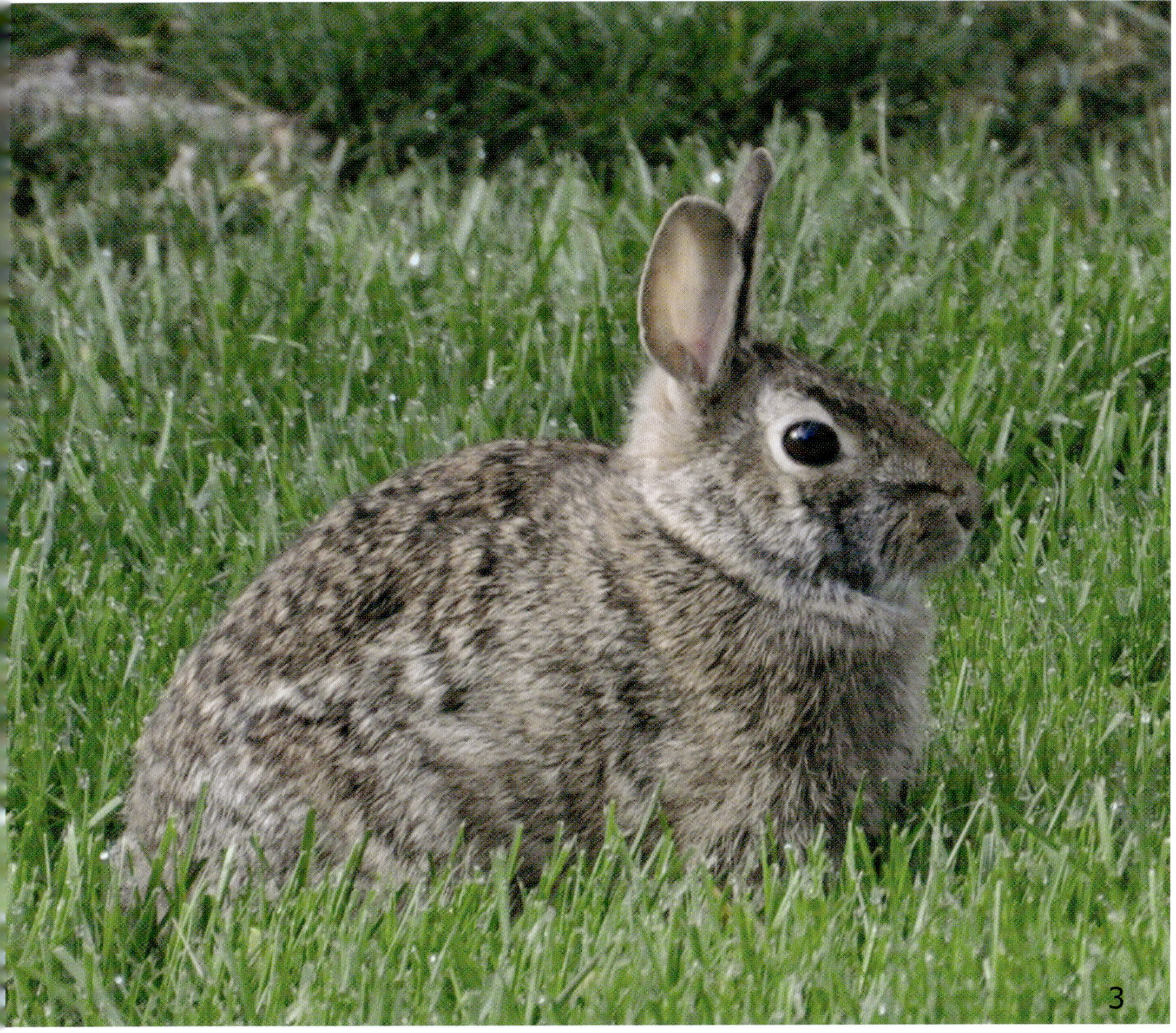

Here is a deer.

The deer is running.

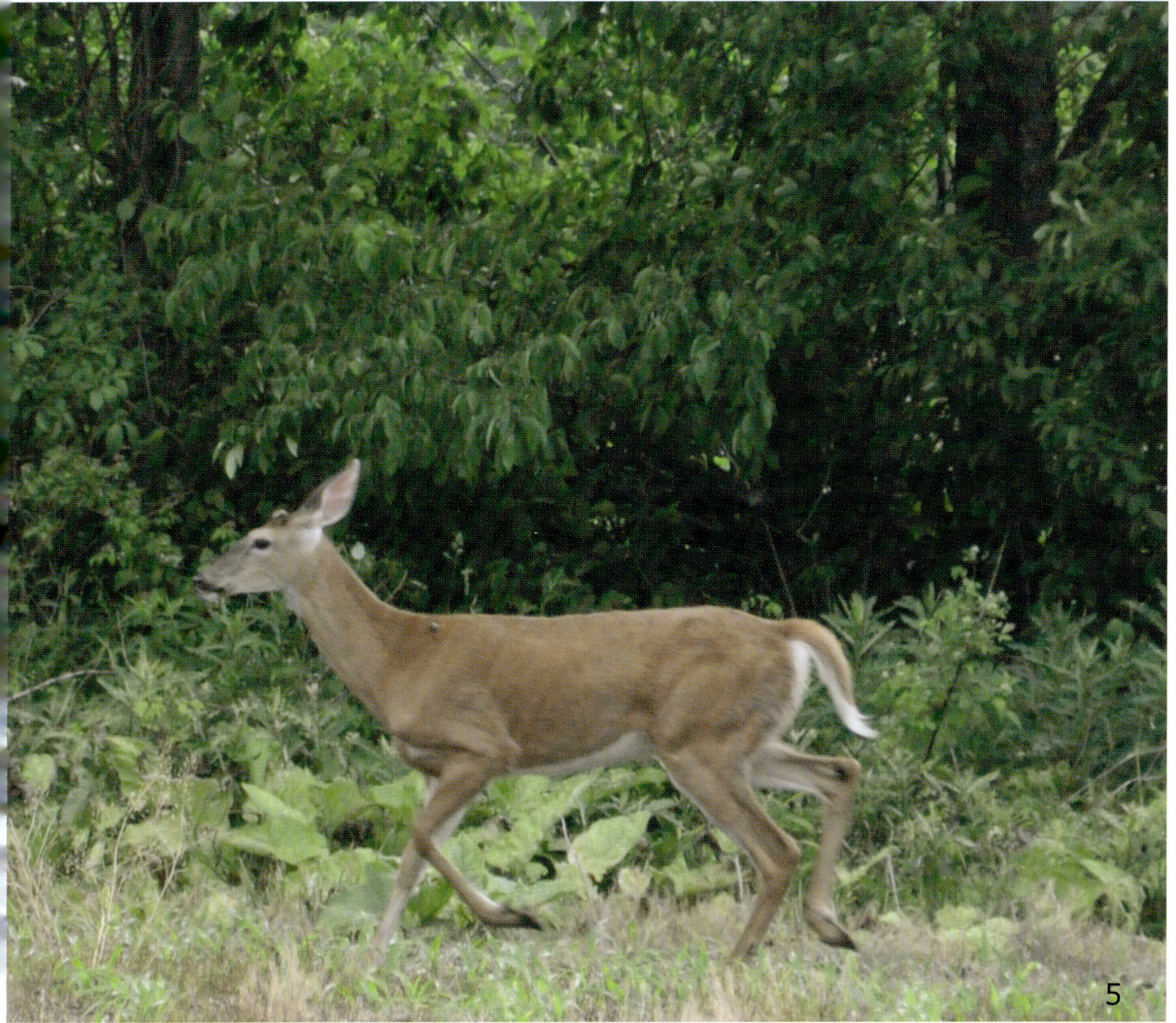

The turkeys are running, too.

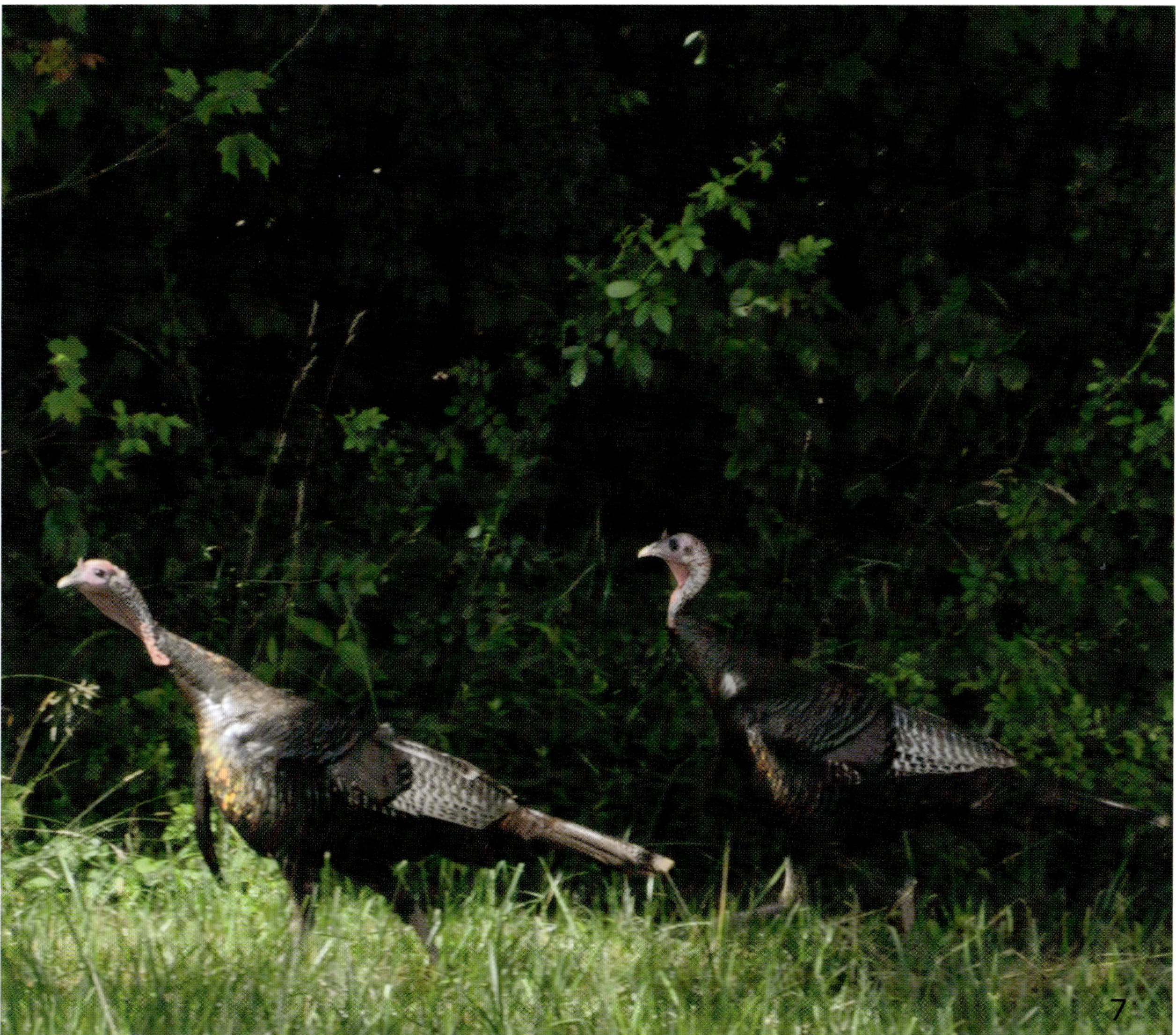

Here is a raccoon.

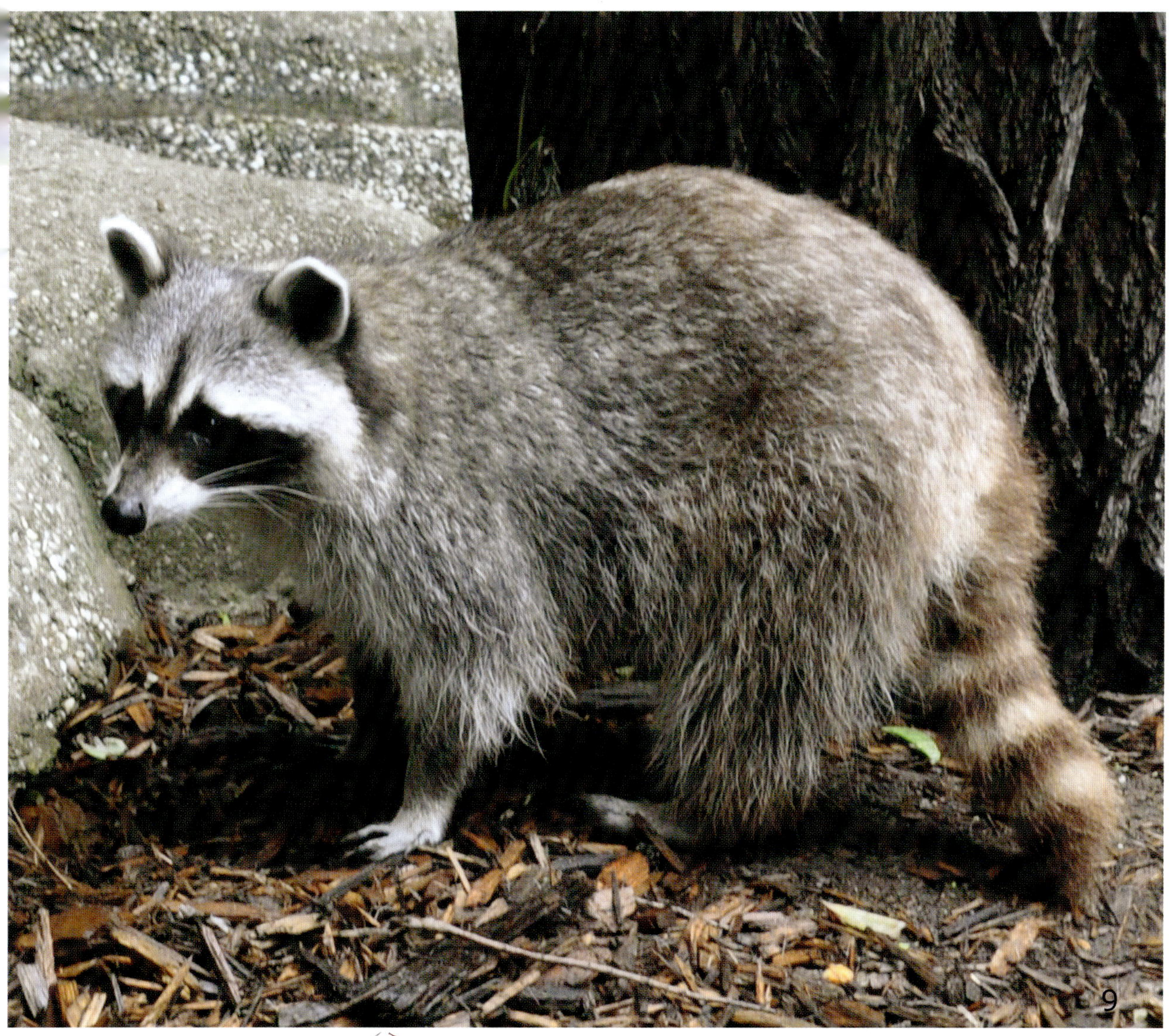

Here is a squirrel.

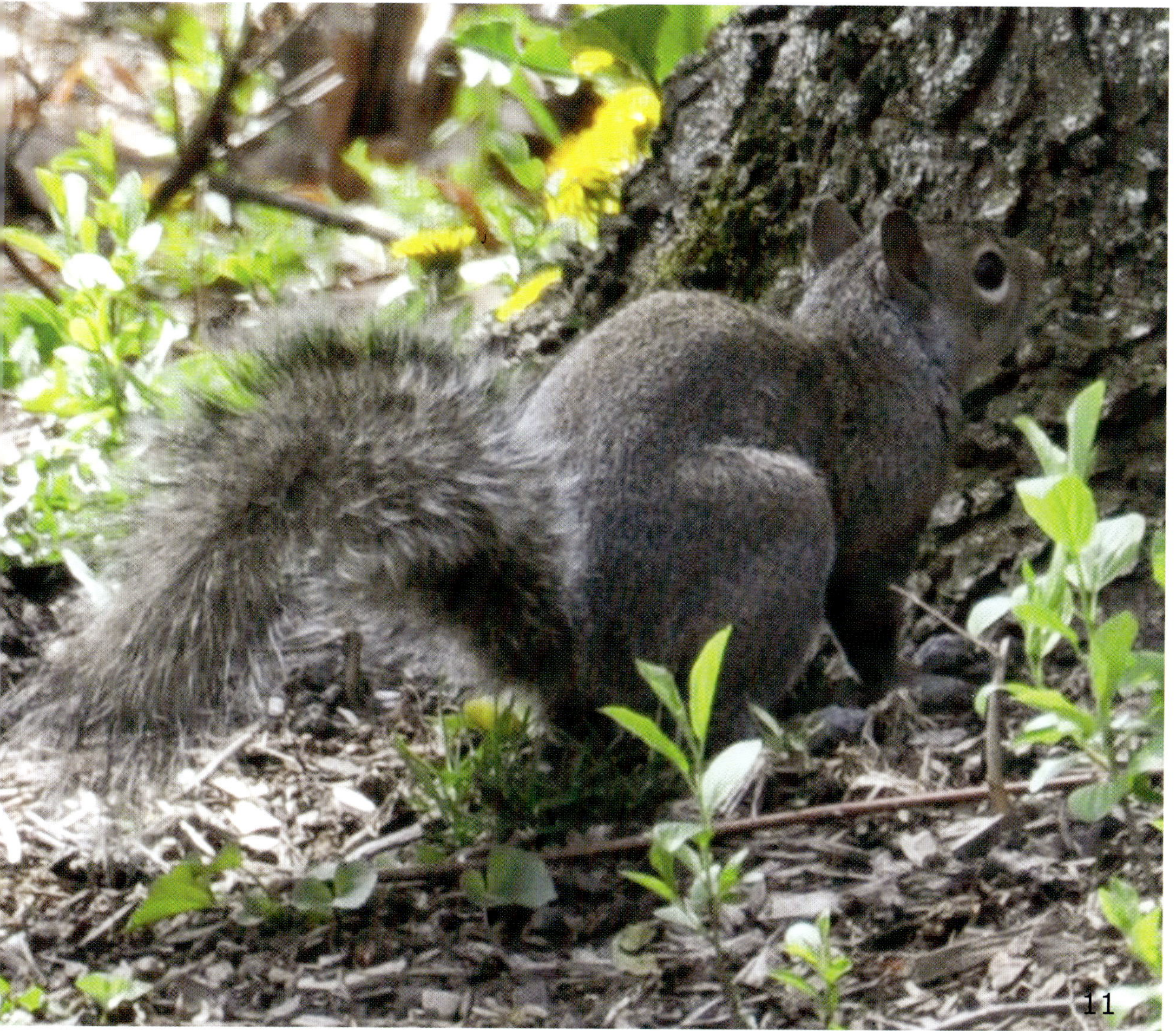

Here is a coyote in the woods.

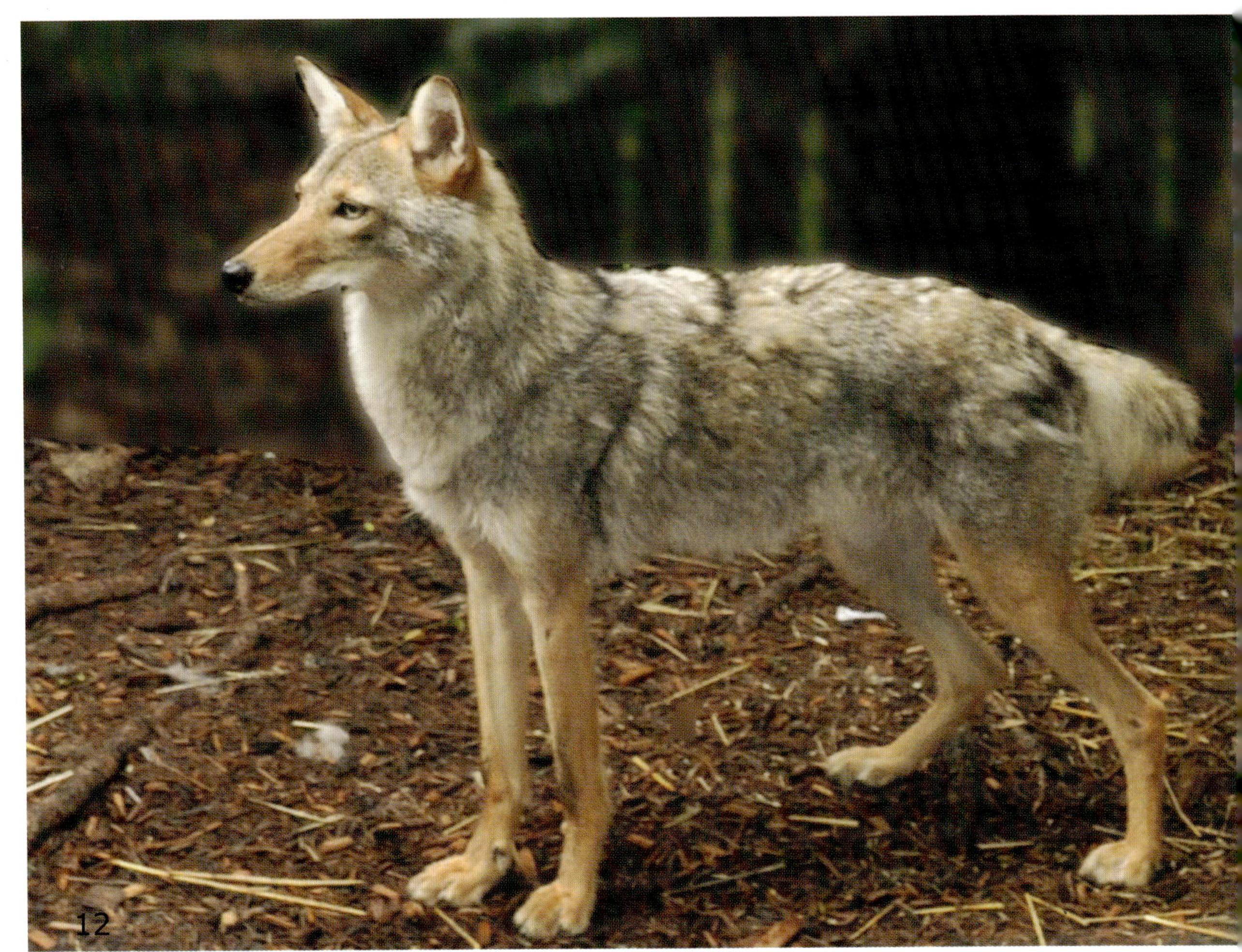